Numerical methods

Unit guide

The School Mathematics Project

CAMBRIDGE
UNIVERSITY PRESS

Main authors	John Higgo
	Paul Roder
	David Tall
	Thelma Wilson
	With contributions from
	Rajiv Bobal
	Stan Dolan
	Kevin Williamson
Team leaders	Paul Roder and David Tall
Project director	Stan Dolan

The authors would like to give special thanks to Ann White for her help in preparing this book for publication.

Graphics by Bunny Graphics

PUBLISHED BY THE PRESS SYNDICATE OF THE UNIVERSITY OF CAMBRIDGE
The Pitt Building, Trumpington Street, Cambridge CB2 1RP, United Kingdom

CAMBRIDGE UNIVERSITY PRESS
The Edinburgh Building, Cambridge CB2 2RU, United Kingdom
40 West 20th Street, New York, NY 10011–4211, USA
10 Stamford Road, Oakleigh, Melbourne 3166, Australia
1 The Moorings, Victoria and Alfred Waterfront, Cape Town 8001, South Africa

First published 1992
Reprinted 1996   2087/11

Produced by 16-19 Mathematics, Southampton

Printed in the United Kingdom at the University Press, Cambridge

ISBN 0 521 42659 6

Contents

Introduction to the unit
(for the teacher)

The aim of this unit is to investigate numerical methods using the computer, to make students aware of possible inaccuracies and of ways to improve the accuracy of numerical methods and minimise the amount of calculation required. The unit begins by considering the inaccuracies that occur in practice and revises techniques developed in *Mathematical methods* to calculate the maximum error that might occur for the given accuracy of the numbers used. It then tackles the problems that arise with computer arithmetic because of the way numbers are stored and processed.

Calculus theory and numerical practice are both used to develop better methods for various calculations, including finding numerical solutions of equations, numerical derivatives, numerical areas and numerical solutions of differential equations. In every case there is an underlying pattern. The calculus is based on the idea of local straightness. This means that under high magnification a small part of a graph, whose function can be differentiated, will look virtually straight. Thus, over a tiny interval, such a graph can be approximated by a straight line. Over a longer distance, however, the graph is likely to look much more curved, and it may be appropriate to use a more curved approximation, such as a quadratic, cubic or higher order polynomial. The basis for this theory is Taylor's theorem, which uses the calculus to give a polynomial approximation to a function near a given point. The numerical methods discussed use various techniques to obtain higher order approximations to produce results more quickly and more accurately.

Chapter 1

This chapter looks at why errors are inherent in the way computers and calculators store numbers. The arithmetic of errors, already covered in the unit *Mathematical methods*, is revised. The need for binary notation is discussed and the principle of floating point notation is introduced. Students should be able to appreciate why even computers can give very inaccurate results and should understand the need for caution in accepting the accuracy of the display of a computer or calculator.

Chapter 2

This chapter builds on the numerical techniques for integration met in *Introductory calculus* and introduces the important concept of the order of accuracy of an approximation. This forms the theme of the whole unit.

Chapter 3

This chapter examines the order of accuracy of two methods for numerical differentiation. These methods are linked to Taylor approximations, which are then developed as the main content of the chapter. Students may have already met Taylor's first approximation and Maclaurin's series in *Calculus methods*. Although the two chapters complement each other, it is not necessary for students to have covered the material in *Calculus methods*.

Chapter 4

The theme of this chapter is solving equations by numerical methods. Students will have met the idea of iteration in *Foundations* and will benefit from having done the extension tasksheet on 'cobweb' and 'staircase' diagrams (Chapter 5, Tasksheet 3E). Although the section on the Newton-Raphson method is self-contained, students will find this much more straightforward if they have previously studied the section on Newton-Raphson in *Calculus methods*.

Chapter 5

This final chapter looks at different methods for solving differential equations and builds on the work covered in *Mathematical methods*. It continues the theme of the order of accuracy of numerical methods and makes a clear link with the methods for numerical integration looked at in Chapter 2.

Graphic calculators/computers

Students are asked to write short programs at certain points in the text. Examples of appropriate programs can be found at the back of this unit guide.

The software package, *Numerical solutions to equations*, has been specifically written for 16-19 Mathematics and is particularly relevant to Chapter 4. Alternatively, students can use graphic calculators. Appropriate programs for iteration (including cobweb and staircase diagrams) and for the Newton-Raphson method can be found in the various packs of graphic calculator copymasters available as part of the *Teacher's resource file* for the main A level course. Short programs for bisection and decimal search are given in this unit guide.

The programs for numerical integration and for solving differential equations have been written in such a way that it should be apparent to students that the two processes are identical as far as the computer is concerned. It should be noted that Simpson's rule is equivalent to the fourth order Runge-Kutta method when each pair of strips in Simpson's rule is regarded as a single strip (i.e. a weighted average of the mid-point and the trapezium rules.)

Tasksheets

5   Differential equations

1 Errors

1.1 Catastrophic errors

> **(a)** Investigate the values of $\dfrac{\sin\left(\frac{\pi}{3}+h\right)-\sin\left(\frac{\pi}{3}\right)}{h}$ for small values of h. A calculator often gives more accurate values than those shown in the table. However, the results may be strange when h is very small.
>
> **(b)** Calculate $\dfrac{e^{h}-e^{0}}{h}$ for small values of h. What happens as h becomes smaller?

(a)   What happens will depend on the make and model of the computer/calculator used. The expected limit as h gets smaller is $\cos\left(\frac{\pi}{3}\right)=0.5$. The author's calculator gave the following results:

h	$\left(\sin\left(\frac{\pi}{3}+h\right)-\sin\left(\frac{\pi}{3}\right)\right)\frac{1}{h}$
1×10^{-5}	0.49999567
1×10^{-6}	0.4999996
1×10^{-7}	0.5
⋮	⋮
1×10^{-12}	0.5
1×10^{-13}	0
1×10^{-14}	0

$\left.\begin{array}{c} \\ \\ \end{array}\right\}$ The value of $\sin\left(\frac{\pi}{3}+h\right)-\sin\left(\frac{\pi}{3}\right)$ is rounded to zero for these values of h.

It is interesting to investigate more closely what happens between 1×10^{-12} and 1×10^{-13}.

h	$\left(\sin\left(\frac{\pi}{3}+h\right)-\sin\left(\frac{\pi}{3}\right)\right)\frac{1}{h}$
1×10^{-12}	0.5
9×10^{-13}	0.5555 …
8×10^{-13}	0.625
7×10^{-13}	0.7142 …
6×10^{-13}	0.8333 …
5×10^{-13}	1
4×10^{-13}	0

What happens is a function of the way this particular calculator stores numbers. Other calculators/computers are likely to give similarly strange results.

(b)   If $y=e^{x}$ then $\dfrac{dy}{dx}=e^{x}$ so you would expect the value of $\dfrac{e^{h}-e^{0}}{h}$ to get closer to $e^{0}=1$ as the value of h gets smaller. On the author's calculator, this does happen until h becomes very small. The calculator then produced strange results for values of h between $h=7.5\times10^{-12}$ and $h=8\times10^{-12}$. For smaller values of h, $e^{h}-e^{0}$ is calculated as zero giving the incorrect result, $\dfrac{e^{h}-e^{0}}{h}\approx0$.

9

The arithmetic of errors

1. $107.5 \pm 0.2 + 6.2 \pm 0.1 = 113.7 \pm 0.3$ cm.
   $107.5 \pm 0.2 - 6.2 \pm 0.1 = 101.3 \pm 0.3$ cm.

2. $12.3 \pm 0.1 + 13.7 \pm 0.1 - 10.4 \pm 0.1 = 15.6 \pm 0.3$

   The least possible value is $\quad\quad 12.2 + 13.6 - 10.5 = 15.3$
   and the greatest possible value is $\quad 12.4 + 13.8 - 10.3 = 15.9$

3. $5.2 \pm 0.05 \approx 5.2\,(1 \pm 0.0096)$
   $3.7 \pm 0.05 \approx 3.7\,(1 \pm 0.0135)$

   The relative error of the product is approximately $0.0096 + 0.0135 = 0.0231$

   $$(5.2 \pm 0.05)\,(3.7 \pm 0.05) \approx 19.24\,(1 \pm 0.0231)$$
   $$\approx 19.24 \pm 0.44$$

   This compares well with the actual maximum value of $5.25 \times 3.75 = 19.6875$
   and the actual minimum value of $5.15 \times 3.65 = 18.7975$.

4. $\dfrac{3.2 \pm 0.05}{1.3 \pm 0.05} \approx \dfrac{3.2\,(1 \pm 0.016)}{1.3\,(1 \pm 0.038)} \approx 2.46\,(1 \pm 0.054) \approx 2.46 \pm 0.13$

   By direct calculation, the maxinumum is $\dfrac{3.25}{1.25} = 2.6$ and the minimum is $\dfrac{3.15}{1.35} \approx 2.33$.

   These compare well because the relative errors are small.

5. If rs and s^2 are so small that they can be ignored,

   $$\frac{(1+r)}{(1-s)} = \frac{(1+r)(1+s)}{(1-s)(1+s)} = \frac{1+r+s+rs}{1-s^2} \approx 1+r+s,$$

   $$\frac{(1-r)}{(1+s)} = \frac{(1-r)(1-s)}{(1+s)(1-s)} = \frac{1-r-s+rs}{1-s^2} \approx 1-(r+s)$$

   $\dfrac{a\,(1 \pm r)}{b\,(1 \pm s)}$ has maximum value $\dfrac{a}{b} \times \dfrac{(1+r)}{(1-s)} \approx \dfrac{a}{b}\,(1+(r+s))$

   and minimum value $\dfrac{a}{b} \times \dfrac{(1-r)}{(1+s)} \approx \dfrac{a}{b}\,(1-(r+s))$

   Therefore, $\dfrac{a\,(1 \pm r)}{b\,(1 \pm s)} \approx \dfrac{a}{b}\,(1 \pm (r+s))$.

Tutorial sheet

1. $100 \times (\pm 0.05) = \pm 5$.

   Usually you would expect accuracy to improve as the number of strips increases. However, if a **very** large number of strips is used then the accumulation of a large number of very small errors can, in fact, make the final calculation less accurate.

2. $5.65 \pm 0.05 \approx 5.65\,(1 \pm 0.00885)$
   $3.2 \pm 0.1 = 3.2\,(1 \pm 0.03125)$

   $$\text{Speed} = \frac{5.65 \pm 0.05}{3.2 \pm 0.1} \approx \frac{5.65\,(1 \pm 0.00885)}{3.2\,(1 \pm 0.03125)}$$

   $$\approx 1.77\,(1 \pm 0.0401)$$

   $$\approx 1.77 \pm 0.07 \text{ ms}^{-1}$$

3. (a) 0.11

   (b) 0.1001100110011 ...

   (c) 0.011011011011 ...

   (d) 0.1010101010 ...

4. The binary expansion of e is 10.101101111110000101 ...

   In floating point form to 10 significant digits, this is 0.1010110111×2^2.

5. (a) This is a fraction where the denominator is a power of 2 and so it terminates.

   (b) This terminates, $0.0625 = \dfrac{1}{2^4}$.

   (c) Although this is a rational fraction ($0.8 = \dfrac{4}{5}$), the denominator **cannot** be written as a power of 2 and so it has a repeating pattern.

   (d) This has a repeating pattern.

   (e) This is an irrational number and so it continues with **no** repeating pattern.

6. (a) This should give exactly 100 as 1 can be stored exactly.

   (b) This should give exactly 100 as $0.5 = \dfrac{1}{2}$ can be stored exactly.

   (c) This may not give exactly 100 as 0.1 cannot be stored exactly and the small errors may accumulate.

   (d) This should give exactly 100 as $0.0625 = \dfrac{1}{2^4}$ and so can be stored exactly.

(continued)

7.　(a)　(i)　$0.000101 = 0.101 \times 2^{-3}$;　 [+ | 1 0 1 | – | 0 1 1]

　　　　(ii)　$10.101 = 0.10101 \times 2^{2}$;　 [+ | 1 0 1 | + | 0 1 0]

　　　　(iii)　$0.0101101 = 0.101101 \times 2^{-1}$;　 [+ | 1 0 1 | – | 0 0 1]

　　(b)　$A = 0.110 \times 2^{-1} = 0.011$

　　　　$B = 0.100 \times 2^{1} = 1.00$

　　　　(i)　$A + A = 0.11 = 0.110 \times 2^{0}$;　 [+ | 1 1 0 | + | 0 0 0]

　　　　(ii)　$A + B = 1.011 \approx 0.101 \times 2^{1}$;　 [+ | 1 0 1 | + | 0 0 1]

　　(c)　$(A + A) = 0.11$ and $B = 1.00$

　　　　$\Rightarrow (A + A) + B = 1.11 = 0.111 \times 2^{1}$

　　　　So $(A + A) + B$ is stored as　 [+ | 1 1 1 | + | 0 0 1]

　　　　$(A + B) = 1.01$ and $A = 0.011$

　　　　$\Rightarrow (A + B) + A = 1.101 \approx 0.110 \times 2^{1}$

　　　　So $(A + B) + A$ is stored as　 [+ | 1 1 0 | + | 0 0 1]

　　　　The two results should be equal but are calculated as being different because $(A + B)$ is stored inaccurately.

2   *Areas*

2.1   Rules for estimating areas under graphs

> You should have noticed that the errors for the trapezium rule
> and the mid-ordinate rule were not the same. By looking at the
> respective errors, suggest a *weighted average* of the formulas for
> the two rules which it would be sensible to try.
>
> Use your data from Tasksheet 1 to investigate the error of the
> answers given by this weighted average as you increase the
> number of strips. What is the order of approximation of this
> new rule?

The errors from the mid-ordinate rule are very close to half of those from the trapezium
rule, so the weighted average should give twice as much weight to the mid-ordinate
rule. It will therefore take the form:

$$\frac{(2 \times \text{mid-ordinate rule answer}) + (\text{trapezium rule answer})}{3}$$

The errors from this formula are very small. It is therefore necessary to work to a large
number of significant figures to see the effect of increasing the number of strips. The
errors are:

	$y = x^4$	$y = x^5$	$y = \cos x$	$y = e^x$
10 strips	2.67×10^{-5}	1.33×10^{-4}	5.06×10^{-7}	3.54×10^{-6}
20 strips	1.67×10^{-6}	8.33×10^{-6}	3.15×10^{-8}	2.24×10^{-7}
30 strips	3.29×10^{-7}	1.64×10^{-6}	6.17×10^{-9}	4.1×10^{-8}
40 strips	1.04×10^{-7}	5.23×10^{-7}	1.84×10^{-9}	1.4×10^{-8}

Halving the strip width approximately divides the error by 16, dividing the width by 3
approximately divides the error by 81, ... The order of appproximation is four.

Accuracy

1. The first and last ordinate rules only give exact answers for constant functions.

2. (a) The trapezium rule answer is too large and the mid-ordinate answer too small.

   (b) The mid-ordinate and trapezium rules give exact answers for constant and linear functions.

   (c) The first ordinate rule gives the area of rectangle *ABEF*.
   The last ordinate rule gives the area of rectangle *ABCD*.

   The trapezium rule area is shaded and is clearly half-way between the other two. So the trapezium rule answer is the mean of the answers given by the first and last ordinate rules.

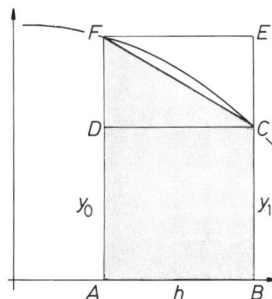

3. The answers given here were obtained using a programmable calculator. Since the first and last ordinate rules are not very accurate, giving answers to 3 d.p. is sufficient. For the other rules the areas are rounded off to 8 d.p. and the errors to 7 d.p. (except where the calculator gave fewer figures), in order to help later work.

$$y = x^4$$

No. of strips	First ordinate		Last ordinate	
	Area	Error	Area	Error
10	4.907	−1.493	8.107	1.707
20	5.627	−0.773	7.227	0.827
30	5.879	−0.521	6.945	0.545
40	6.007	−0.393	6.807	0.407

No. of strips	Mid-ordinate		Trapezium	
	Area	Error	Area	Error
10	6.34676	−0.05324	6.50656	0.10656
20	6.3866725	−0.0133275	6.42666	0.02666
30	6.39407523	−0.0059248	6.41185054	0.0118505
40	6.39666703	−0.0033330	6.40666625	0.0066663

(continued)

14

4. (a)
$$y = x^5$$

No. of strips	First ordinate		Last ordinate	
	Area	Error	Area	Error
10	7.733	−2.934	14.133	3.466
20	9.133	−1.533	12.333	1.667
30	9.630	−1.037	11.763	1.096
40	9.883	−0.783	11.483	0.817

No. of strips	Mid-ordinate		Trapezium	
	Area	Error	Area	Error
10	10.5338	−0.1328667	10.9328	0.2661333
20	10.6333625	−0.0333042	10.7333	0.0666333
30	10.65185761	−0.0148091	10.69628971	0.0296230
40	10.65833516	−0.0083315	10.68333125	0.0166646

(b)
$$y = \cos x$$

No. of strips	First ordinate		Last ordinate	
	Area	Error	Area	Error
10	1.048	0.139	0.765	−0.145
20	0.979	0.070	0.838	−0.072
30	0.956	0.047	0.862	−0.048
40	0.945	0.035	0.874	−0.036

No. of strips	Mid-ordinate		Trapezium	
	Area	Error	Area	Error
10	0.91081469	0.0015173	0.90626441	−0.0030330
20	0.90967641	0.0003790	0.90853955	−0.0007579
30	0.90946584	0.0001684	0.90896063	−0.0003368
40	0.90939215	0.0000947	0.90910798	−0.0001894

(c)
$$y = e^x$$

No. of strips	First ordinate		Last ordinate	
	Area	Error	Area	Error
10	5.771	−0.618	7.049	0.660
20	6.075	−0.314	6.714	0.325
30	6.178	−0.211	6.604	0.215
40	6.231	−0.158	6.550	0.161

No. of strips	Mid-ordinate		Trapezium	
	Area	Error	Area	Error
10	6.37842008	−0.0106360	6.41033877	0.0212827
20	6.38639477	−0.0026613	6.39437943	0.0053233
30	6.38787309	−0.0011830	6.39142224	0.0023661
40	6.38839062	−0.0006655	6.39038710	0.0013310

5. −

1. $$I = \left[\tfrac{1}{3}ax^3 + \tfrac{1}{2}bx^2 + cx \right]_{-h}^{h}$$

$$I = \left(\tfrac{1}{3}ah^3 + \tfrac{1}{2}bh^2 + ch \right) - \left(-\tfrac{1}{3}ah^3 + \tfrac{1}{2}bh^2 - ch \right) = \tfrac{2}{3}ah^3 + 2ch$$

2. $$y_1 = a \times 0^2 + b \times 0 + c = c$$

$$y_2 = ah^2 + bh + c$$

3. $$\tfrac{1}{3}h(y_0 + 4y_1 + y_2) = \tfrac{1}{3}h[(ah^2 - bh + c) + 4c + (ah^2 + bh + c)]$$

$$= \tfrac{1}{3}h(2ah^2 + 6c)$$

$$= \tfrac{2}{3}ah^3 + 2ch.$$

Cubics

1. B is $(0, d)$
   C is $(h, ah^3 + bh^2 + ch + d)$

2. The equation for B is $r = d$
   The equation for C is $ph^2 + qh + r = ah^3 + bh^2 + ch + d$

3. Since $r = d$:

   $$ph^2 - qh = -ah^3 + bh^2 - ch \quad \text{①}$$
   $$\text{and} \quad ph^2 + qh = ah^3 + bh^2 + ch \quad \text{②}$$

   Adding ① and ② gives:

   $$2ph^2 = 2bh^2, \text{ so } p = b$$

   Subtracting ① from ② :

   $$2qh = 2ah^3 + 2ch$$
   $$\Rightarrow \quad q = ah^2 + c$$

   So $y = px^2 + qx + r$ becomes
   $y = bx^2 + (ah^2 + c)\, x + d$

4. $$\text{LHS} - \text{RHS} = \int_{-h}^{h} (ax^3 - ah^2x)\, dx$$

   This is the integral of an odd function from $-h$ to h and so is zero.

1.

No. of strips	Trapezium rule error	Simpson's rule error
10	−0.03707	−0.0146
20	−0.01313	−0.00515
40	−0.004645	−0.00182
100	−0.001176	−0.000459
200	−0.000416	−0.000162

2.   (a)   Each angle at the centre of the circle is 60°.
Using the formula $\frac{1}{2}\,ab \sin c$,
the area of one triangle is $\frac{1}{2}$ x 4 sin 60°.
So the area of $1\frac{1}{2}$ triangles is 3 sin 60°.

    (b)   The area of one triangle is
$\frac{1}{2}$ x 4 sin 45°
There are 2 triangles in a quarter-
circle, so these have total area
4 sin 45°

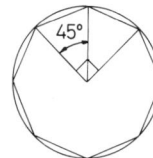

    (c)   A regular dodecagon will have 3 triangles in a quarter-circle and each will have
area 2 sin 30°. So 6 sin 30° < π.
For an *n*-sided polygon, the inequality is $\frac{1}{2}\,n \times \sin\!\left(\frac{360°}{n}\right) < \pi$.

3.   The program merely needs to output the value of $\frac{1}{2}\,n \times \sin\!\left(\frac{360°}{n}\right)$ for different values
of *n*.

n	No. of triangles in quarter circle	Error in estimate for π
40	10	−0.01290
80	20	−0.00323
160	40	−0.000807
400	100	−0.000129
800	200	−0.0000323

Notice that, as the number of divisions increases, the accuracy of this method improves
faster than that of Simpson's rule. This method has second order of accuracy.

4.   Simpson's rule is not very accurate because the graph of $y = \sqrt{(4 - x^2)}$ near $x = 2$ is not
at all like a cubic. The gradient of the graph is infinite at that point.

(continued)

5.   (a)   For $\frac{1}{\cos x}$, Simpson's rule quickly gives 1.22619 (to 5 d.p.).

For $\frac{1}{\sin x}$ and $\frac{1}{\tan x}$, a programmable calculator gives the same answers (to the nearest integer), for 10, 20, 30, 40, 100 and 200 strips. They are as follows:

10 strips	33336
20 strips	16670
40 strips	8338
100 strips	3339
200 strips	1673

The graph of $y = \frac{1}{\cos x}$ is reasonably similar to a cubic near $x = 0$. However, for both $y = \frac{1}{\sin x}$ and $y = \frac{1}{\tan x}$, y tends to infinity as x approaches 0.

(b)   It is not possible to integrate $y = \frac{1}{\sin x}$ and $y = \frac{1}{\tan x}$ from $x = 0$ to $x = 1$ because y is undefined when $x = 0$.

6.   By algebraic integration, $\int_0^1 x^{1/2} \, dx = 0.\dot{6}.$ Simpson's rule does not give this answer very quickly; using 50 strips the value given is 0.6664, to 4 d.p. The problem here is the infinite gradient at $x = 0$.

$\int_0^1 x^{-2} \, dx$ is undefined.

By algebraic integration, $\int_{0.01}^1 x^{-2} \, dx = 99.$ Simpson's rule with 100 strips gives 102.6

Similarly, $\int_{0.001}^1 x^{-2} \, dx = 999.$ Simpson's rule with 1000 strips gives 1036.

$y = x^{-2}$ causes more problems with Simpson's rule because y tends to infinity as x approaches 0.

3 Taylor polynomials

3.1 Quadratic approximations

> **(a)** Explain why $f(a) + f'(a)h + f''(a)\dfrac{h^2}{2}$ is likely to be a good approximation to $f(a+h)$ for small values of h.
>
> **(b)** Explain why $(2+h)^{-1} \approx \dfrac{1}{2} - \dfrac{h}{4} + \dfrac{h^2}{8}$ for small values of h.
>
> **(c)** Graph $y = x^{-1}$ and $y = \dfrac{1}{2} - \dfrac{(x-2)}{4} + \dfrac{(x-2)^2}{8}$ on the same axes.
>
> **(d)** Explain why $\ln(1+h) \approx h - \dfrac{h^2}{2}$ for small values of h.
>
> **(e)** Rewrite the approximation given in (d) in terms of x and y and check your answer on a graph plotter.

(a) If $g(x)$ is chosen so that $g(a) = f(a)$, $g'(a) = f'(a)$ and $g''(a) = f''(a)$, then

$$g(a+h) = f(a) + f'(a)h + f''(a)\frac{h^2}{2}$$

$f(x)$ will be very similar to $g(x)$ near $x = a$ and so

$$f(a+h) \approx f(a) + f'(a)h + f''(a)\frac{h^2}{2}$$

(b) $f(x) = x^{-1} \Rightarrow f(2) = \dfrac{1}{2}$

$f'(x) = -x^{-2} \Rightarrow f'(2) = -\dfrac{1}{4}$

$f''(x) = 2x^{-3} \Rightarrow f''(2) = \dfrac{1}{4}$

$f(2+h) = (2+h)^{-1} \approx \dfrac{1}{2} + \left(-\dfrac{1}{4}\right)h + \left(\dfrac{1}{4}\right)\dfrac{h^2}{2}$

$\approx \dfrac{1}{2} - \dfrac{1}{4}h + \dfrac{1}{8}h^2$

(c) As was to be expected, the graphs look very similar near $x = 2$.

(d) $f(x) = \ln x \Rightarrow f(1) = 0$

$f'(x) = x^{-1} \Rightarrow f'(1) = 1$

$f''(x) = -x^{-2} \Rightarrow f''(1) = -1$

$\ln(1+h) \approx 0 + 1 \times h - 1 \times \dfrac{h^2}{2}$, for small h

$\approx h - \dfrac{h^2}{2}$

(e) $y = \ln(1+x) \approx x - \dfrac{x^2}{2}$

The graphs are similar near $x = 0$.

Investigating radius of convergence

1.   The series converges for $-1 < x < 1$.

2.   Adjust the program by using the fact that the term containing x^{n+1} is obtained from the term containing x^n by multiplying by $\frac{-nx}{n+1}$.

       The series is valid for $-1 < x \le 1$.

3.   Even for higher approximations, the series is not valid for $|x| > 1$.

4.   The series should appear to converge for $-1 < x \le 1$.

5.   (a)   The series are valid for all x. As higher degree Taylor approximations are used, the approximation should be seen to be good for an increasing range of x values.

       (b)   For the e^x series, the term in x^n is obtained from the term in x^{n-1} by multiplying by $\frac{x}{n}$.

             If $x = 5$, this ratio is less than 1 for $n \ge 6$ and gets smaller as n increases. The series therefore converges faster than a geometric progression with ratio $\frac{5}{6}$.

1.   (a)
$$\begin{aligned}
f(x) &= \cos x &\Rightarrow\quad f(0) &= 1\\
f'(x) &= -\sin x &\Rightarrow\quad f'(0) &= 0\\
f''(x) &= -\cos x &\Rightarrow\quad f''(0) &= -1\\
f^{(3)}(x) &= \sin x &\Rightarrow\quad f^{(3)}(0) &= 0\\
f^{(4)}(x) &= \cos x &\Rightarrow\quad f^{(4)}(0) &= 1\\
f^{(5)}(x) &= -\sin x &\Rightarrow\quad f^{(5)}(0) &= 0\\
f^{(6)}(x) &= -\cos x &\Rightarrow\quad f^{(6)}(0) &= -1
\end{aligned}$$

$$\cos x \approx 1 + 0x - \frac{x^2}{2!} + 0x^3 + \frac{x^4}{4!} + 0x^5 - \frac{x^6}{6!}$$

$$\cos x = 1 - \frac{x^2}{2!} + \frac{x^4}{4!} - \frac{x^6}{6!} + \dots$$

   (b) The series is valid for all x.

2.   (a) $\sqrt{(1-x)} \approx 1 - \frac{1}{2}x - \frac{1}{8}x^2 - \frac{1}{16}x^3 - \frac{5}{128}x^4$

   (b) Putting $x = 0.02$ gives $\sqrt{0.98} \approx 0.9899494938$.

        This gives $\sqrt{2} \approx 1.414213563$, which is accurate to 9 s.f.

   (c) One method is to use $\sqrt{108} = 6\sqrt{3} = 10\sqrt{1.08}$.

        $\sqrt{1.08} \approx 1.0392304 \Rightarrow \sqrt{3} \approx 1.732050667$, which is accurate to 7 s.f.

3.   (a) $\sqrt{(1+x^2)} \approx 1 + \frac{1}{2}x^2 - \frac{1}{8}x^4 + \frac{1}{16}x^6 - \frac{5}{128}x^8 + \frac{7}{256}x^{10}$

   (b)
$$\int_0^{0.5} (1+x^2)^{\frac{1}{2}}\, dx \approx \left[x + \frac{1}{6}x^3 - \frac{1}{8} \times \frac{x^5}{5} + \frac{1}{16} \times \frac{x^7}{7} - \frac{5}{128} \times \frac{x^9}{9} + \frac{7}{256} \times \frac{x^{11}}{11} \right]_0^{0.5}$$

$$\approx 0.5201145745$$

   (c) Simpson's rule gives 0.5201144097 with both 50 strips and 100 strips.

        Taylor's tenth approximation gives an answer correct to 5 decimal places.

        This might have been a useful method in pre-computer days!

(continued)

4. (a) After the fourth approximation the value for ln 1.1 is 0.09531to 4 s.f.

    (b) After the fourth approximation the value for ln 1.2 is 0.1823 to 4.s.f.

        After the seventh approximation the value for ln 1.4 is 0.3365 to 4 s.f.

        After the thirteenth approximation the value for ln 1.6 is 0.4700 to 4 s.f.

5E. This could make an interesting group project, giving some insight into the problems of compiling tables.

4 *Solving equations*

4.1 Chaos

> **Discuss how to solve each of the following equations.**
>
> (a) $x + 1 = 4$
> (b) $4x - 1 = 3x$
> (c) $x^2 + 2 = 3x$
> (d) $7 - x^5 = 3x$
> (e) $\sin x + e^{x-5} = 0$

The first two equations can easily be solved be rearrangement.

(a) $x = 3$

(b) $x = 1$

(c) $x^2 - 3x + 2 = 0 \Rightarrow (x - 2)(x - 1) = 0 \Rightarrow x = 1$ or 2

(d)   The equation has only one real solution, 1.2628 to 4 d.p.

Decimal search is a possible method.

Iteration, using $x_{n+1} = (7 - 3x_n)^{0.2}$, is an efficient method.

(e)   The equation has an infinite number of solutions and is studied in the next section.

Step by step to chaos

1.

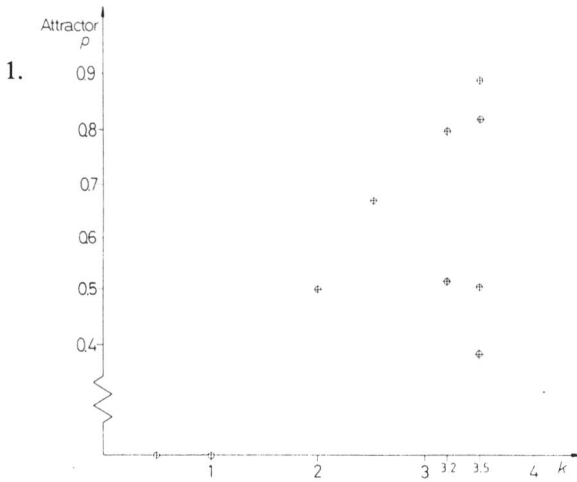

2.   See the program in this unit guide.

3.   Change the line

$$kp(1-p) \rightarrow p$$

to

$$k(p^3 - 3p^2 + 2p) \rightarrow p$$

4.   The following results were obtained using a graphic calculator.

For $kp(1-p)$

$k_1 = 3$, $k_2 = 3.45$, $k_3 = 3.55$ and $k_4 = 3.57$

So $h_1 = 0.45$, $h_2 = 0.1$ and $h_3 = 0.02$

For $k(p^3 - 3p^2 + 2p)$

$k_1 = 1.8$, $k_2 = 2.19$, $k_3 = 2.28$ and $k_4 = 2.3$

So $h_1 = 0.39$, $h_2 = 0.09$ and $h_3 = 0.02$

5.   **For $kp(1-p)$:** $\dfrac{h_1}{h_2} = 4.5$, $\dfrac{h_2}{h_3} = 5$

   For $k(p^3 - 3p^2 + 2p)$: $\dfrac{h_1}{h_2} = 4.3$, $\dfrac{h_2}{h_3} = 4.5$

(continued)

6.   (a) $k_2 = k_1 + h_1 = 3$

$h_2 \approx \dfrac{2}{\delta} \approx 0.4283$

$k_3 = k_2 + h_2 \approx 3.4283$

$h_3 \approx \dfrac{0.4283}{\delta} \approx 0.0917$

$k_4 = k_3 + h_3 \approx 3.5201$

(b) $h_2 \approx \dfrac{2}{\delta} \Rightarrow h_3 \approx \dfrac{2}{\delta^2}$ and so on.

(c) $1 + \dfrac{2}{1 - \dfrac{1}{\delta}}$

≈ 3.5451

(d)   It is approximately the value of k at which chaos starts.

It is interesting to note that, using the values found for k_1 and k_2, chaos should start at

$3 + \dfrac{0.45}{1 - \dfrac{1}{\delta}} \approx 3.57$ for question 2.

and at

$1.8 + \dfrac{0.39}{1 - \dfrac{1}{\delta}} \approx 2.30$ for question 3.

These predictions seem to be accurate.

Bisection and decimal search

It is easier to write a program for bisection than for decimal search. Without a program, it is easier to use decimal search. A program for bisection is given in this unit guide.

1.   (a)   11 steps are needed to be certain the root is 1.260.

         (b)   21 steps are needed to be certain the root is 1.259921

2.   (a)   18 steps are needed to show the solution is between 1.259 and 1.260. Testing 1.2595 shows that the solution is nearer to 1.260. 19 steps are therefore needed in all. (A similar comment applies to all other decimal search answers.)

         (b)   32 steps are needed to show the solution is between 1.259921 and 1.259922.

Bisection is more efficient for this equation – the difference is greater for 6 decimal places than for 3 decimal places.

The answers to questions 3 and 4 assume that the integer bounds for the solutions are first found graphically.

3.   (a)   Using bisection, 11 steps are needed to be certain the root is 2.303.
               Using decimal search, 8 steps are needed to show the solution is between 2.302 and 2.303.

         (b)   Using bisection, 22 steps are needed to be certain the root is 2.302776.
               Using decimal search, 30 steps are needed to show the solution is between 2.302775 and 2.302776.

Because the digits of the answer are small, decimal search is, at first, more efficient. For 6 decimal places, bisection is more efficient.

4.   (a)   Using bisection, 11 steps are needed to be certain the root is 1.609.
               Using decimal search, 17 steps are needed to show the solution is between 1.609 and 1.610.

         (b)   Using bisection, 21 steps are needed to be certain the root is 1.609290.
               Using decimal search, 38 steps are needed to show the solution is between 1.609289 and 1.609290.

Bisection is more efficient in both cases.

In general, bisection will become more efficient as the accuracy is increased, because the interval containing the solution is halved each time and so quickly becomes very small.

Investigating iteration

1.   (a) $(x-1)^2 - 1 - 5 = 0$ or use formula:

$$(x-1)^2 = 6$$
$$x - 1 = \pm \sqrt{6}$$
$$x = 1 \pm \sqrt{6}$$

$$x = \frac{(2 \pm \sqrt{4 + 20})}{2}$$

$$x = \frac{(2 \pm \sqrt{24})}{2}$$

$$x = 1 \pm \sqrt{6}$$

$x = -1$ and $x = 3$ would seem to be reasonable starting points for the iterations in the rest of this question.

    (b)   Starting at $x = -1$ and at $x = 3$ leads to $-1.4495 \approx 1 - \sqrt{6}$

    (c) $g'(1 + \sqrt{6}) \approx -2.38$ $g'(1 - \sqrt{6}) \approx -0.42$

2.   (a)   Another rearrangement is $x = \dfrac{2x + 5}{x}$.

    (b)   For $x = \sqrt{2x + 5}$, iterations starting at $x = -1$ and at $x = 3$ lead quickly to $3.4495 \approx 1 + \sqrt{6}$.

           For $x = -\sqrt{2x + 5}$, iterations starting at -1 lead to -1.4495. Starting at 3 does not work since the iteration requires the square root of a negative number.

           $x = \frac{1}{2}(x^2 - 5)$ produces no useful result – the values either oscillate or diverge.

           $x = \frac{2x + 5}{x}$ leads to 3.4495, when starting at $x = -1$ and when starting at $x = 3$.

    (c)

If $g(x) = \sqrt{2x + 5}$,	$g'(1 + \sqrt{6}) = 0.29$	and	$g'(1 - \sqrt{6}) = 0.69$
If $g(x) = -\sqrt{2x + 5}$,	$g'(1 + \sqrt{6}) = -0.29$	and	$g'(1 - \sqrt{6}) = -0.69$
If $g(x) = \frac{1}{2}(x^2 - 5)$,	$g'(1 + \sqrt{6}) = 3.45$	and	$g'(1 - \sqrt{6}) = -1.45$
If $g(x) = \frac{2x + 5}{x}$,	$g'(1 + \sqrt{6}) = -0.42$	and	$g'(1 - \sqrt{6}) = -2.38$

3.   In the questions above, the iterations only converge to roots at which $|g'(x)| < 1$.

The Newton-Raphson method

1.   (a)   Programs are given for a CASIO graphic calculator.

   (i) $? \rightarrow X$
          Lbl 1
          $X - (X - \cos X) \div (1 + \sin X) \rightarrow X$ ◢
          Goto 1

   (ii) $? \rightarrow X$
          $? \rightarrow H$
          Lbl 1
          $X - \cos X \rightarrow A$
          $X + H \rightarrow X$
          $X - \cos X \rightarrow B$
          $X - H \rightarrow X$
          $X - HA \div (B - A) \rightarrow X$ ◢
          Goto 1

   (iii) $? \rightarrow X$
          $? \rightarrow H$
          Lbl 1
          $X - H \rightarrow X$
          $X - \cos X \rightarrow A$
          $X + 2H \rightarrow X$
          $X - \cos X \rightarrow B$
          $X - H \rightarrow X$
          $X - \cos X \rightarrow Y$
          $X - 2HY \div (B - A) \rightarrow X$ ◢
          Goto 1

(b) and (c).   The efficiency of the numerical methods will depend on the value of 'h' used. In most cases, if $h = 0.001$ (or smaller) the numerical programs will give answers as quickly as the 'algebraic' program. It is necessary to use the more complicated second order method only if the graph is so curved that a **very** small value of h is needed.

2E.   (a) $g'(x) = 1 - \dfrac{f'(x)^2 - f(x) f''(x)}{f'(x)^2} = \dfrac{f'(x)^2 - f'(x)^2 + f(x) f''(x)}{f'(x)^2} = \dfrac{f(x) f''(x)}{f'(x)^2}$

   (b)   (i) $f(x)$ will be small near the root and so, from the above equation, you would expect $g'(x)$ to be small.

          (ii)   If $f'(x)$ is very small near the root or if $f''(x)$ is very large, then, $g'(x)$ may not be very small.

29

1. The approach would be similar to solving equations by bisection. If the dictionary had 300 pages, the teacher could ask if the word was between pages 1 and 150, then between 1 and 75, and so on. After the page was found, a similar method would locate the line containing the word.

    The *Compact Oxford English Dictionary* is claimed to have over half a million words (in very small print) on 2416 pages. Here it would take a maximum of 12 questions to locate the page. If there were up to 256 words to a page, 8 further questions would locate the word. 2416 x 256 is well over half a million, so 20 questions would suffice!

2. (a) For any starting point, the iteration oscillates between the initial and one other number.

    (b) If $y = \frac{2x+1}{x-2}$

    then $y(x-2) = 2x+1$

    $\Rightarrow yx - 2y = 2x + 1$

    $\Rightarrow x(y-2) = 2y + 1$

    $\Rightarrow x = \frac{2y+1}{y-2}$

    So $g^{-1}x = g(x)$

    Students may prefer to show that $gg(x) = x$.

    (c) Since g is self-inverse, two iterations will always return to the first value.

3. (a) It quickly becomes clear that most negative starting points lead to convergence on $x = -2$, while positive values give convergence on $x = 2$. In the latter case, the convergence can be quite rapid at first, but it then slows down.

    (b) If $f(x) = x^3 - 2x^2 - 4x + 8$, $f'(x) = 3x^2 - 4x - 4$
    $= (3x + 2)(x - 2)$

    So there are turning points at $x = -\frac{2}{3}$ and $x = 2$.

    Since there is a turning point at the positive solution, convergence to this point will be relatively slow. All starting points with x values less than $-\frac{2}{3}$ will give convergence to -2 whereas those greater than $-\frac{2}{3}$ will give convergence to $+2$.

(continued)

4.   (a) and (b)

If $0 < k < 1$ the iteration converges on $x = 0$.
The convergence is slower when k is near 1.

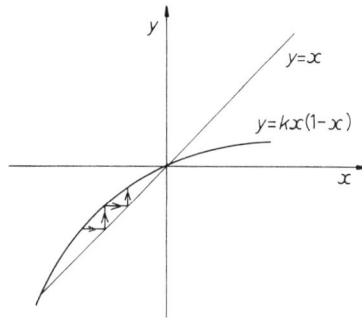

If $1 < k < 2$, the iteration converges on $x = a$ as in the 'staircase' diagram. The convergence is slower when k is near 1.

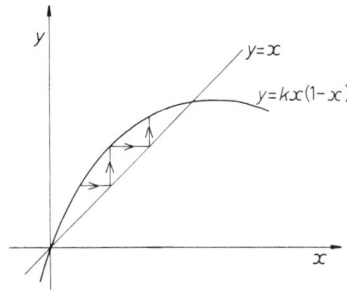

If $2 < k < 3$, the iteration converges on $x = a$ as in the 'cobweb' diagram. The convergence is slower when k is near 3.

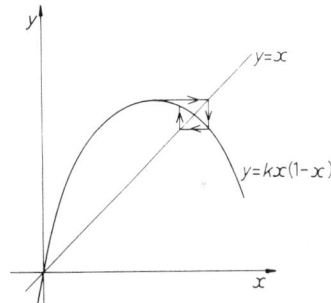

The iteration does not converge when $k > 3$. When $k = 3.2$ the iteration oscillates between two values, when $k = 3.5$ it oscillates between four values and when $k = 4$ it oscillates in a random (chaotic) manner between $x = 0$ and $x = 1$.

(c) $g'(x) = k - 2kx$, so $g'(0) = k$. The iteration will only converge when $| g'(0) | < 1$, and this will be when $k < 1$.

(d)   (i) $ka - ka^2 = a$ and $k - 2ka = -1$, leads to $a = \frac{2}{3}$

        (ii) $k = 3$

        (iii)   When $k = 3$, $| g'(a) | = 1$. If $k > 3$, $| g'(a) | > 1$, so the iteration will not converge.

31

5   *Differential equations*

5.1   Euler's method

Complete the table.

(a)   What happens to the error (approximately) as the
         step length is halved?

(b)   By considering the steps $\frac{1}{2}$ and $\frac{1}{20}$, decide what
         happens when the step length is divided by 10.

(c)   What appears to be the order of approximation of
         Euler's method?

(d)   Make an educated guess as to how many steps might
         be needed to reduce the error to 0.0001. (Do not
         check this on a graphic calculator!)

Step	Estimated value of y when $x = 2$	Error
1/2	5.875	2.125
1/4	6.9063	1.0938
1/8	7.4453	0.5547
1/16	7.7207	0.2793
1/20	7.7763	0.2238

(a)   The error is approximately halved.

(b)   The error is approximately divided by 10.

(c)   Euler's method appears to be a first order approximation.

(d)
$$\begin{array}{ccc} & \text{Step} & \text{Error} \\ \div\,2238 \ \Big(& \begin{array}{c} 0.05 \\ 0.00002234 \end{array} & \begin{array}{c} 0.2238 \\ 0.0001 \end{array} \Big) \div 2238 \end{array}$$

The error is 0.2238 when the step length is $\frac{1}{20}$. If the method is a first order
approximation then the error will be 0.0001 when the step length is $\frac{1}{20}$ divided
by 2238 i.e. 0.00002. This, however, would require 50 000 steps! While a
powerful computer might achieve this degree of accuracy in a reasonable amount
of time, a graphic calculator might take about 30 minutes using the program given
in this unit guide. More efficient methods are therefore needed to obtain reasonably accurate results.

1. If the step length is divided by a factor λ, then the error is divided by λ.

2. If the step length is divided by a factor λ, then the error is divided by λ^2.

3. Euler's method appears to be a first order approximation. The improved Euler and mid-point Euler methods both appear to be second order approximations.

    For Euler's method, reducing the step from $\frac{1}{4}$ to $\frac{1}{40}$ reduces the error from 1.625 to 0.174 and not to 0.1625. Euler's method is a first order approximation, but only for 'small' values of h.

4. (a)

Step	Euler's method	Improved Euler	Mid-point Euler
1/2	0.4945	0.00295	0.00148
1/4	0.2480	0.00074	0.00037
1/20	0.0497	0.000029	0.000015
1/40	0.0249	0.0000074	0.0000037

    (b)

Step	Euler's method	Improved Euler	Mid-point Euler
1/2	22.41	4.3938	2.1698
1/4	12.29	1.1120	0.5543
1/20	2.64	0.0447	0.0223
1/40	1.33	0.0112	0.0056

Both of these results suggest (for small step lengths) that Euler's method is a first order approximation and that the other two methods are second order approximations. Notice also that for small step lengths the error of the mid-point method is consistently half that of the improved Euler method.

The weighted average

1. $$y_{n+1} = \frac{2(y_n + hg(x_n + \frac{h}{2})) + (y_n + h\frac{g(x_n) + g(x_n + h)}{2})}{3}$$

$$= \frac{4y_n + 4hg(x_n + \frac{h}{2}) + 2y_n + h(g(x_n) + g(x_n + h))}{6}$$

$$= y_n + h\frac{g(x_n) + 4g(x_n + \frac{h}{2}) + g(x_n + h)}{6}$$

2. See the program *Fourth order Runge-Kutta* in this unit guide.

3.
Step	Error
1	0
1/2	0
1/4	0
1/8	0

The weighted average of the mid-point Euler and the improved Euler methods is called the **fourth order Runge-Kutta** method. This gives a precise result for the differential equation $\frac{dy}{dx} = 3x^2$.

4. (a) The method gives precise results for quadratic functions.

   (b) The method gives precise results for cubic functions.

   (c) An example: $\frac{dy}{dx} = x^4$ with $x = 1$, $y = 1$ has an algebraic solution $y = 0.2x^5 + 0.8$ and so $y = 7.2$ when $x = 2$. The fourth order Runge-Kutta method gives

Step	Error
1	8.33×10^{-3} $\left.\right\}\div 2^4$
1/2	5.21×10^{-4} $\left.\right\}\div 2^4$
1/4	3.26×10^{-5} $\left.\right\}\div 2^4$
1/8	2.03×10^{-6}

   (d) The algebraic solution is $y = \sin x$ and so $y = \sin 2$ when $x = 2$.

Step	Error
1	3.25×10^{-4}
1/2	1.99×10^{-5}
1/4	1.24×10^{-6}
1/8	7.71×10^{-8}

When the step length is halved, the error is divided by approximately 2^4. The method is a fourth order approximation.

1.

Step	y
0.1	1.88447
0.05	1.85095
0.01	1.82570

2. See the program *Extended Improved Euler* in this unit guide.

3.

Step	y
0.1	1.820153843
0.05	1.819757906
0.01	1.819599377

4. (a) $y = k e^{-\frac{x^2}{2}}$

Let $u = -\dfrac{x^2}{2}$, then $\dfrac{du}{dx} = -x$

$y = ke^{u} \implies \dfrac{dy}{du} = ke^{u} = y$

$\dfrac{dy}{dx} = \dfrac{du}{dx} \times \dfrac{dy}{du} = -xy$

(b) If $\dfrac{dy}{dx} = -xy$, then $y = ke^{-\frac{x^2}{2}}$.

If $y = 3$ when $x = 0$, then $y = 3e^{-\frac{x^2}{2}}$.

(c) $y = 3e^{-0.5} \approx 1.819591979$

5. **Euler's method**

	Step	Error	
$\div 2$	0.1	0.06488	$\div 2.07$
$\div 5$	0.05	0.03136	$\div 5.13$
	0.01	0.00611	

The method appears to be first order.

Improved Euler's method

	Step	Error	
$\div 2$	0.1	0.000562	$\div 3.4 = 2^{1.77}$
$\div 5$	0.05	0.000166	$\div 22.4 \approx 5^{1.93}$
	0.01	0.00000740	

The method appears to be second order.

6. Use Euler's method to calculate an estimate of y at $(x_n + \frac{h}{2})$ and calculate the gradient at this mid-point. Then use this gradient to move from (x_n, y_n) to (x_{n+1}, y_{n+1}).

(continued)

7.

	Step	y	Error	
$\div 2$	0.1	1.817962086	0.00163	$\div 2^{2.05}$
	0.05	1.819199168	0.000393	
$\div 5$	0.01	1.819576709	0.0000153	$\div 5^{2.02}$

This appears to be a second order method. Notice that with this extended method, the error of the mid-point Euler method is **not** consistently half that of the improved Euler method. In fact, it is greater. The extended fourth order Runge-Kutta method is not, therefore, a simple weighted average of the two methods. (The method can be improved by using the improved Euler method to obtain the mid-point estimate of y. However, the new method is still of second order.)

8.　　　　The solution curves all have the familiar 'bell' shape of the Normal curve.

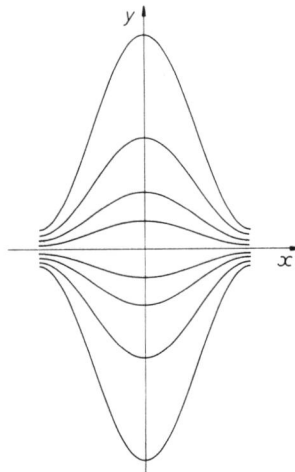

1. (a)

Step	Estimate value of y when $x = 1$	Error
1/2	95.648	96.154
1/4	95.437	95.943
1/10	2.640	3.147
1/20	−2.115	1.609

   (b) The problem here is that $\frac{dy}{dx}$ and y both oscillate at a greater rate than the step size (period $= \frac{2\pi}{100} \approx 0.06$). The method can be seen to be of second order when the step lengths are smaller than 0.01.

2. (a) If you divide the step length by λ then you divide the error by approximately λ^2 for a second order approximation and by approximately λ^4 for a fourth order approximation.

   (b)

	Step	Error	
÷ 5 (	0.2	0.0125	) ÷ 5^4
	0.04	**0.00002**	

3. (a) $y - 3.906 = 5^2 \times (y - 4.002)$

   $\Rightarrow \quad y = 4.006$

   (b)

	Step	Error	
÷ 20 (	−0.1	0.004	) ÷ 400 = 20^2
	−0.005	0.00001	

4. (a)

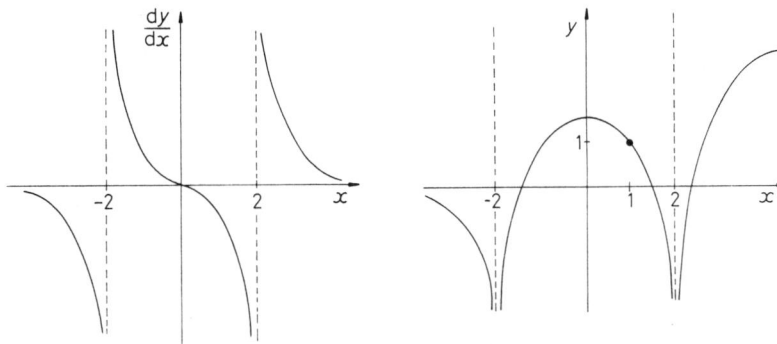

   (b) Starting at (1, 1) you could calculate y for any value of x in the range $-2 < x < 2$. A numerical method might jump over the asymptote at $x = 2$ to give you a solution at $x = 3$, but the solution would not be valid.

Programs

The following programs have been written for a CASIO *fx-7000G* calculator. Only minor changes will be needed to make them compatible with other makes of calculator (they should work on other CASIO models without alteration).

The main changes needed to adapt the programs for a TEXAS TI–81 (or TI–85) graphic calculator are as follows:

CASIO *fx-7000G*	TEXAS TI–81	
"X" ? $\rightarrow$ X	Disp "X"	
	INPUT X	
"AREA" : A ◢	Disp "AREA"	
	Disp A	
	Pause	(not needed at the end of a program)
Prog 1	Prgm 1	
N > 0 $\Rightarrow$ Goto 1	If N > 0	
	Goto 1	
Plot (X, Y)	PT-On (X, Y)	

Many of the programs use an infinite loop. For these programs, press $\boxed{\text{AC}}$ to stop the program.

For programs which display a graph, set the range before you start and use $\boxed{\text{G}\leftrightarrow\text{T}}$ to view the graph.

The programs are given under their appropriate chapter headings.

The following program for numerical integration evaluates the integral $\int_A^B f(x)\, dx$ by dividing the area under the curve into N strips of width H.

"A" ? → A
"B" ? → B
"N" ? → N
0 → S
(B − A) ÷ N → H
A → X   Replace with A + H → X for the last ordinate rule
Lbl 1
Prog 1
N − 1 → N   Replace with N − 2 → N for Simpson's rule
N > 0 ⇒ Goto 1
"AREA" : S

The 'Prog 1' statement diverts the program to a subroutine which depends on the numerical method used for evaluating the integral. (You can type in the subroutine **in place of** the 'Prog 1' statement if you wish, but having it as a separate routine makes it easier for you to edit the program to use different methods.)

You should use the following subroutines for the methods indicated. The subroutine is stored as a program by the calculator. The precise program number used is immaterial, but number 1 has been assumed in the main program.

First/last ordinate rule	Mid-ordinate rule	Trapezium rule	Simpson's rule
***** → Y	X + H ÷ 2 → X	***** → P	***** → P
S + HY → S	***** → Y	X + H → X	X + H → X
X + H → X	S + HY → S	***** → Q	***** → Q
	X + H ÷ 2 → X	(P + Q) ÷ 2 → Y	X + H → X
		S + HY → S	***** → R
			P + 4Q + R → Y
			S + HY ÷ 3 → S

In each of the above, replace ***** with the function being integrated. For example, write $\sin(X^2)$ in place of ***** if the integral is $\int \sin x^2 \, dx$.

The following program will enable you to investigate the radius of convergence of the Taylor series

$$\frac{1}{1-h} = 1 + h + h^2 + h^3 + \ldots$$

by evaluating the partial sum $S_r = 1 + h + h^2 + h^3 + \ldots + h^r$ for different values of r and plotting these on a graph.

It is important to use h rather than x as the calculator plots all points as (x, y) coordinates and so the value of x would not remain constant but would increase as r increases.

Radius of convergence

```
    ? → H
    0 → R▲
    0 → S
→   Lbl 1
|   S + H^R → S▲        Change this line for a different Taylor series.
↑   Plot R, S▲          (Note that H^R will appear as Hx^y R)
|   R + 1 → R▲
◄── Goto 1
```

Graph when $h = 0.6$

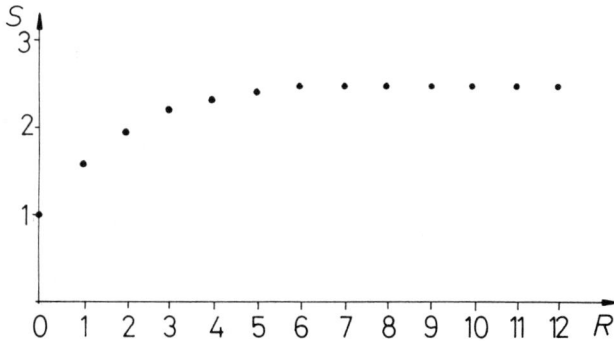

The programs on this page will help you investigate Robert May's population model, based on the iterative formula

$$p_{n+1} = kp_n(1 - p_n)$$

Iteration diagram

This program enables you to draw a graph of a sequence as shown below while obtaining the corresponding numerical values.

Range 0, 50, 10, 0, 1, 0.1

"P" ? → P
"K" ?→ K
0→ N
Lbl 1
N + 1→ N▲
K P (1 − P) → P▲
Plot N, P▲
Goto 1

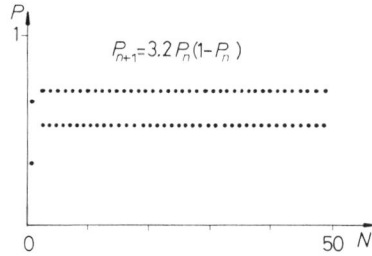

Bifurcation diagram

This program plots the attractors of the iteration formula for successive values of K, thus producing a bifurcation diagram. (See Tasksheet 1E – *Step by step to chaos*)

Range 0, 4, 1, 0, 1, 0.1   (1)

0.2 → S
0 → K   (2)
Lbl 1
S → P
1 → N
Lbl 2
K P (1 − P)→ P
N ≥ 60 ⇒ Plot K, P
N + 1→ N
N ≤ 80 ⇒ Goto 2
K + 0.1→ K   (3)
K ≤ 4 ⇒ Goto 1   (4)

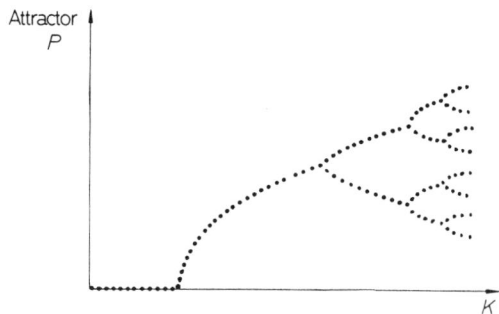

More detailed investigation of the diagram can be obtained by changing the range statement (1) to zoom in on part of the diagram and by changing the lines marked (2), (3) and (4).

Approximate measurements can be obtained from a plotted diagram by plotting a point within the range of the screen using the 'Plot' command (e.g. Plot 3, 0.5) and **then** using the arrow keys to move the point around the screen to take readings.

The following programs are for solving equations of the form $f(x) = 0$ by the bisection method and by the decimal search method.

Bisection

        $? \rightarrow X$
        $1 \rightarrow H$
        Prog 1
        $Y > 0 \Rightarrow$ Goto 2
        Lbl 1
        Prog 1   The program assumes that integer bounds are
        X ◢   known and that the initial input of X is the
        Y ◢   lower bound.
        $Y < 0 \Rightarrow X + 0.5H \rightarrow X$
        $Y > 0 \Rightarrow X - 0.5H \rightarrow X$
        $0.5H \rightarrow H$
        Goto 1
        Lbl 2
        Prog 1
        X ◢
        Y ◢
        $Y > 0 \Rightarrow X + 0.5H \rightarrow X$
        $Y < 0 \Rightarrow X - 0.5H \rightarrow X$
        $0.5H \rightarrow H$
        Goto 2

The 'Prog 1' statement diverts the program to a subroutine which contains the function whose zero you are finding. For example, Prog 1 would contain $X^3 - 2 \rightarrow Y$ if the equation is $x^3 - 2 = 0$. The precise program number is immaterial, but number 1 has been assumed in the main program.

Decimal search

A fully automatic program is rather lengthy and so the following short program is suggested.

        $? \rightarrow X$
        $? \rightarrow H$
        Lbl 1
        $X + H \rightarrow X$
        $***** \rightarrow Y$
        X ◢
        Y ◢
        Goto 1

Stop the program when the sign of Y changes (e.g. from positive to negative), then run the program again, but input the new X value (i.e. the new lower bound) and the new value of H.

The following programs are for solving differential equations of the form $\frac{dy}{dx} = f(x)$.

Use this program if you wish to see the x and y values at each step.	Use this program if you wish to specify the number of steps, N, and display only the solution.
" X " ? → X " Y " ? → Y " H " ? → H Lbl 1 Prog 1 X ◢ Y ◢ Goto 1	" X " ? → X " Y " ? → Y " H " ? → H " N " ? → N Lbl 1 Prog 1 N − 1 → N N > 0 ⇒ Goto 1 X ◢ Y

The 'Prog 1' statement diverts the program to a subroutine which depends on the numerical method used for solving the differential equation. (You can type in the subroutine **in place of** the 'Prog 1' statement if you wish, but having it as a separate routine makes it easier for you to edit the program to use different methods.)

You should use the following subroutines for the methods indicated. The subroutine is stored as a program by the calculator. The precise program number used is immaterial, but number 1 has been assumed in the main program.

Euler's method	Mid-point Euler	Improved Euler	Fourth order Runge-Kutta
***** → G Y + HG → Y X + H → X	X + H ÷ 2 → X ***** → G Y + HG → Y X + H ÷ 2 → X	***** → A X + H → X ***** → B (A + B) ÷ 2 → G Y + HG → Y	***** → A X + H ÷ 2 → X ***** → B X + H ÷ 2 → X ***** → C (A + 4B + C) ÷ 6 → G Y + HG → Y

In each of the above, replace ***** with the right-hand side of the differential equation. For example, write $\sin(X^2)$ in place of ***** if $\frac{dy}{dx} = \sin x^2$.

The differential equations programs can be adjusted to display the x and y values at each step of the solution and plot these on the calculator's graphic display, thus giving you an image of the solution curve. To do this, insert the following three lines immediately after the 'Prog 1' statement in the main program.

$$X \to U : Y \to V$$
$$\text{Plot } U, V$$
$$U \to X : V \to Y$$

Many differential equations are functions of both x **and** y. Below are two subroutines which extend the mid-point Euler and the improved Euler methods to deal with such equations. (Euler's method needs no adjustment.)

Extended Mid-point Euler

$$***** \to G$$
$$Y \to S$$
$$Y + HG \div 2 \to Y$$
$$X + H \div 2 \to X$$
$$***** \to G$$
$$S + HG \to Y$$
$$X + H \div 2 \to X$$

This saves the initial value of y while the mid-point value is estimated using Euler's method.

Extended Improved Euler

$$***** \to A$$
$$Y \to S$$
$$X + HA \to Y$$
$$X + H \to X$$
$$***** \to B$$
$$(A + B) \div 2 \to G$$
$$S + HG \to Y$$

This saves the initial value of y while an interim value is calculated using Euler's method.

In each of the above replace ***** with the right-hand side of the differential equation. For example, write $-XY$ in place of ***** if $\frac{dy}{dx} = -xy$.

Both routines will, of course, work if the differential equation is in terms of x only.